Cybersecurity
PARENTS LLC

For permissions or licensing inquiries, please contact:
Cybersecurity Parents LLC
www.cybersecurityparents.com
info@cybersecurityparents.com

First edition, May 2025
ISBN: 979-8-218-67472-4

The author generated images in part with GPT-3, OpenAI's large-scale language-generation model. Upon generating draft images, the author reviewed, edited, and revised the content to their own liking and takes ultimate responsibility for the content of the images in this publication.

Legal Disclaimer

This book is intended for informational and educational purposes only. The views expressed are those of the authors and do not constitute legal, professional, or technical advice. Readers should consult appropriate professionals or official sources for guidance specific to their circumstances.

All product names, logos, brands, and trademarks mentioned in this book are the property of their respective owners. These names are used for identification purposes only, and their use does not imply endorsement, affiliation, or sponsorship.

While every effort has been made to ensure the accuracy of the information contained herein, the publisher and author make no representations or warranties with respect to the completeness or accuracy of the contents and disclaim any liability in connection with the use of this information.

Cybersecurity Parents LLC and its representatives shall not be held liable for any damages or loss arising from the use of the material in this book.

Want to share this book with a family member or friend?

Direct them to our website by scanning the code below for purchasing options and our contact information:

cybersecurityparents.com

Introduction

The internet has reshaped how we live, learn, and connect –
but for parents, it has also reshaped the world in which we
raise our children. Cyberbullying, identity theft, online
predators, and data breaches are just some of the digital
threats children face today, with real-world consequences.

This book provides clear, practical guidance to help parents
protect their children's privacy and security online.
Whether you're safeguarding your child's digital footprint
from birth, navigating their first steps online, or managing a
teen immersed in social media and gaming, you'll find
essential strategies to keep them safe.

We begin by breaking down the key cybersecurity risks
kids and teens face. From there, we offer actionable steps –
from securing devices to teaching responsible online habits
– so you can protect your family's private information and
respond effectively to cyber threats.

Technology and threats will continue to evolve, but with
the right knowledge, you can empower your children to
protect themselves in our digital world.

Thank you for being curious and helping the next
generation.
– Chad Rychlewski and Kae David

Who Are Cybersecurity Parents?

We are experienced cybersecurity professionals who protect organizations from online threats. We apply this expertise to safeguarding our own families.

Our goal is to equip you with the tools and knowledge to protect your children in an increasingly complex digital world. We hope this book serves as a practical guide you can return to as your family's online habits evolve.

Chad Rychlewski is a cybersecurity leader with over 16 years of experience helping organizations and families stay safe in an ever-changing digital world. Chad's passion for cybersecurity education grew from years of working at the intersection of technology, risk, and real-life parenting concerns. When he's not helping families stay cyber-safe, he's speaking, writing, and leading initiatives to make cybersecurity more accessible for everyone. His goal is to give parents the knowledge and confidence they need to guide their children through today's connected world.

With over a decade of experience in cybersecurity, Kae David is a recognized leader in protecting both individuals and enterprises from the evolving threats of the digital world. Driven by a passion for security and a deep belief that informed users are the first line of defense, Kae has dedicated her career to making complex security concepts accessible and actionable.

Chapter 1

WHY
CYBERSECURITY
MATTERS
FOR PARENTS

Introduction

"Parents spend an average of just 46 minutes discussing online safety throughout their child's upbringing."
- *Kaspersky's Global Research Report (2019)*

Before your child takes their first steps, they may have already left a digital one.

From the moment a photo is posted or a name is entered into an online form, a lasting trace begins – one that can shape your child's future in ways you might not expect. In this chapter, we'll explore what a digital footprint really is and why it matters more than ever in today's tech-driven world. You'll learn the difference between active and passive data collection, the hidden risks behind everyday actions like school logins and social media posts, and how digital exposure begins far earlier than most parents realize.

We'll also take a hard look at who really controls your child's digital presence – from tech companies and school systems to well-meaning relatives. You'll uncover the platforms and spaces where children spend most of their time online, along with the dangers that come with each.

By the end of this chapter, you'll not only understand the scope and permanence of your child's digital footprint, but also recognize the urgency in taking a more active role in protecting it. This is the first step toward empowering yourself – and your child – in the digital age.

What is a Digital Footprint?

The moment your child's image, name, or personal details are posted online, their digital footprint begins. Unlike footprints in the sand that disappear with the tide, digital footprints are permanent, trackable, and often out of your control. They grow with every online interaction – whether it's a birth announcement on Facebook, a preschool photo on Instagram, or a school login on an educational platform.

Once your child enters this world, their privacy can be at risk in the digital one. When they enter preschool or elementary school, this risk multiplies from online applications, portals, picture release forms, and more. Their Social Security Number becomes a target for data breaches.

There are two types of digital footprints:

- **Active Footprint** – Information intentionally shared online, such as creating a TikTok video, signing up for a gaming account, or posting a comment on YouTube.

- **Passive Footprint** – Data collected behind the scenes without direct input, such as browsing habits tracked by websites, location data collected by apps, or behavior monitored by third-party cookies.

Every time your child's name is entered into an online form, image is uploaded to social media, or device connects to a website, their digital footprint grows. The real concern isn't just what's being shared, but who has access to it and how it's being used.

Who Controls Your Child's Digital Footprint?

As a parent, you play a major role in shaping your child's online presence, often before they even understand what the internet is. However, just like the real world, the control you have over their digital footprint eventually extends far beyond your hands.

Who else contributes to your child's digital presence?

- **Tech Companies** – Social media platforms, search engines, and online services collect and store data, often using it for advertising, artificial intelligence (AI) training, or even selling it to third parties.

- **Schools & Educational Platforms** – Many schools use digital learning tools that collect student information. While some claim to have strong security measures, data breaches have happened.

- **Friends, Family & Strangers** – Well-meaning relatives might post photos or share personal details about your child without considering privacy risks. Even a simple birthday shoutout with their full name can expose more than intended.

The problem? Once data is online, you have little control over where it spreads or how it's used. A single shared image can be copied, stored, and redistributed without your consent. Even secure platforms are vulnerable, leading to personal data breaches by the likes of hackers.

For example, your child's school will have a Student Information System (SIS) containing personal data such as their name, contact information, date of birth, limited medical alert information, and Social Security Number. One such system was hacked in 2024, exposing over 70 million records.[1] As a parent, this is a frightening reality.

The Digital Landscape: Where Kids Are Online

Children today are digital natives, growing up in a world where online interactions are a daily norm. Most schools provide personal computers or tablets as part of their curriculum. As children explore these tools, they begin accessing websites and applications that may lack proper restrictions.

From phones, tablets, and personal computers to smart TVs – if it can connect to the internet, it likely has a game or application that enables online interaction. Here's where kids spend most of their time and some of the risks associated with each:

- **Social Media**
  Risks: Overexposure, privacy invasion, cyberbullying, and interactions with strangers.

- **Online Gaming**
  Risks: Cyberbullying, phishing scams, grooming by predators, and financial exploitation.

---

[1] *Education Week, 2024*

- **Educational Tools & Virtual Learning**
  Risks: Data collection, tracking, and device compromise.

- **Streaming & Entertainment**
  Risks: Targeted ads, inappropriate content, and addiction to algorithm-driven video feeds.

- **AI Chatbots & Virtual Assistants**
  Risks: Misinformation, inappropriate content, and reduced critical thinking.

- **Online Marketplaces & Shopping Apps**
  Risks: Accidental purchases, scams, and access to age-inappropriate products.

- **Anonymous & Encrypted Chat Apps**
  Risks: Unmonitored conversations, explicit content, and potential exploitation.

Shocking Stats on Children's Internet Exposure

To put the digital reality into perspective, here are some alarming statistics:

- **89% of U.S. school districts** provided each student with access to their own device starting in **Grade 1** with access to the internet.[2]

- **40% of children ages 8-12** have a social media account.[3]

---

[2] Education Week Research Center, 2021

[3] Marketing Dive, 2023

- **46% of teenagers ages 13-17** have experienced some form of cyberbullying.[4]

- **63% of tweens and 77% of teens** have encountered inappropriate content online.[5]

- Parents spend an **average of just 46 minutes** discussing online safety throughout their child's upbringing.[6]

- **81.25% of children's apps** contain trackers, and **4.47% request location permissions.**[7]

- In **2023, there were 121 ransomware attacks** on U.S. K-12 schools.[8]

These numbers highlight how quickly kids are integrated into the online world – often without the knowledge or maturity to navigate it safely.

Conclusion

In today's hyper-connected world, every parent must take an active role in managing their child's digital footprint. You are their first line of defense, shaping their online identity before they fully understand its consequences.

Now that you have a foundational understanding of how digital footprints are created, who controls them, and why they matter, the next step is learning how to navigate social

---

[4] Avast, 2023

[5] Bark Annual Report, 2024

[6] Kaspersky, 2019

[7] arxiv.org, 2023

[8] Comparitech, 2024

media safely and teaching your children to make smart choices about what they share online.

Next Steps for Parents

This can feel like a heavy topic, but **small steps add up.** Start today with these **simple actions** to reduce risks for your child.

Step 1: Start Conversations Early *(5 Minutes)*

*Near the end of this book there is a section called **Talking Points**. Utilize the different topics by age group to start talking to your children.*

- **Talk to your child** about online privacy in an age-appropriate way.
- **Teach them simple rules**, like never sharing their real name, address, or school online.

Bonus: Make it a **habit** to discuss online safety casually – during meals, car rides, or bedtime chats.

Step 2: Think Before You Post *(5 Minutes)*

- **Pause before sharing** photos or details about your child online.
- **Consider the long-term impact** – what you post today may follow them into adulthood.

Bonus: Do a **privacy sweep** of your social media – remove old posts that might reveal personal details. This can include information like birthdays, home address, phone numbers, or frequent places you visit.

Step 3: Encourage Safe Habits *(5 Minutes)*

- **Set basic guidelines** for internet use, e.g., no screens in bedrooms at night.
- **Model good digital behavior** – show them how you handle privacy and security online.

Bonus: Try setting aside your devices when your kids get home from school and model **less device-dependent behaviors.**

Step 4: Freeze Your Child's Credit
(Currently, this is a manual and lengthy process for minors which involves mailing in a paper form, including copies of official identifier documents, per agency.)

- **Protect their identity** by freezing their credit with **Experian, Equifax, and TransUnion.**
- **Try it for yourself first** – it's much easier for adults, only takes a few minutes per credit bureau online!

Bonus: By doing this step, you help prevent someone from opening credit cards, loans, or more using your child's information as part of identity theft.

Chapter 2

UNDERSTANDING THE RISKS FOR KIDS & TEENS IN CYBERSECURITY

Introduction

Technology is woven into nearly every part of our lives – and our children's lives are no exception. From lullabies played on smart speakers to homework submitted through classroom apps, digital tools are present from the very beginning. But with this convenience comes a new layer of responsibility for parents: understanding the different stages of tech exposure and how to guide our children safely through each one. Just as we help our kids learn to walk, talk, and read, we must also help them develop healthy digital habits that will stay with them for life.

Each age and stage comes with its own unique risks and opportunities. A toddler may not be chatting online, but their digital footprint can begin with a single photo shared on social media. An elementary schooler might be exploring educational games, but could stumble onto inappropriate content without proper safeguards. As they grow, tweens and teens start building their own digital identities, engaging on platforms where peer pressure, privacy concerns, and online interactions carry real-world consequences. Understanding what's appropriate – and what's risky – at every step helps parents provide age-appropriate guidance without fear-based restrictions.

This chapter will walk you through the key stages of a child's tech journey, highlighting what to watch for, what to teach, and how to stay involved in ways that build trust and confidence. You'll also learn about four common types of online risks and how to respond to each with practical strategies. The goal isn't to control everything, but to equip your child (and yourself) to handle the digital world with awareness, courage, and connection. Let's explore how you can be a steady guide through the ever-changing world of technology.

The Different "Stages" of Technology with Your Children

Technology exposure evolves as children grow and so do the associated risks. Here's what to consider at each stage:

- **Infancy & Toddler Years (Ages 0-5):** This stage is crucial for parents in managing their child's digital footprint.
  - Avoid sharing too much personal information or images of your child online, as these can contribute to a permanent digital record.
  - Be mindful of privacy settings when posting on social media and avoid geotagging photos.
  - Limit screen exposure and focus on hands-on learning rather than digital engagement.

- **Kindergarten - 5th Grade (Ages 5-10):** This stage sets the foundation of digital literacy you and your children will be able to expand upon as they continue to grow their digital footprint.
  - Introduce basic tech literacy.
  - Use child-friendly apps and monitor usage closely.

- Ensure your children only access age-appropriate content.
- Begin teaching them the basics of online privacy, such as not talking to strangers online.
- Reinforce that not everything they see online is real or safe.

- **Middle School (Ages 11-13):** This is often when kids have their first experiences with social media and connected platforms, such as multiplayer games.

  - Set clear rules for online interactions.
  - Help them understand the importance of not sharing personal details, the risks of engaging with strangers, and how to recognize suspicious behavior online.
  - Encourage them to talk to you about any uncomfortable interactions.

- **High School Years (Ages 14-18):** Independence increases, while your visibility into their online habits may decrease. However, this is when they have a higher likelihood of exposure to cyber threats.

  - Have honest discussions about privacy, digital reputation, and online risks.
  - Encourage them to think critically about what they post and who they engage with online.

      ○ Teach them how to handle online conflicts, privacy settings, and the long-term impact of their digital footprint.

4 Common Risks to be Aware of

Risk 1: Social Media Dangers

Social media platforms are designed to be engaging, but they come with privacy risks. Children and teens often overshare personal details, such as their location, school, or daily routines, making them vulnerable to predators and identity theft.

Exposure to inappropriate content and online predators is another concern, as is peer pressure, which can lead to risky behaviors like participating in dangerous viral challenges or sexting. The addictive nature of social media can also negatively impact mental health.

How to Navigate:

- **Set profiles to private** and ensure your child only connects with real-life friends.
- **Teach your child** to never share real-time locations, personal addresses, or any identifiable information publicly.
- **Review privacy settings together** and discuss the long-term consequences of oversharing.
- **Encourage taking breaks** from social media to support mental well-being.

Bonus: Review your settings on a monthly basis to ensure nothing has changed or talk with your child and decide if there is a need for something to be updated.

Risk 2: Gaming & Chat Apps

Online gaming and chat apps expose children to unfiltered conversations where strangers can interact with them freely. Risks include cyberbullying, exposure to inappropriate discussions, online grooming by predators, and financial risks associated with in-game purchases.

Some platforms also allow voice chat, increasing the likelihood of encountering offensive or harmful interactions. Additionally, excessive gaming can contribute to addiction and social isolation.

How to Navigate:

- **Enable parental controls** and restrict chat functions to prevent interactions with unknown individuals.
- **Set up a separate email** for gaming accounts to avoid phishing scams.
- **Educate your child** on never sharing passwords, personal information, or financial details.
- **Encourage a healthy balance** between gaming and offline activities.
- **Keep computer or gaming systems in communal areas** so you have visibility into what your child is doing.

Bonus: Have your child tell you what types of online safety tips they should follow while engaging on these gaming & chat apps.

Risk 3: Phishing & Scams – How Children Get Tricked Online

Scammers target children through fake giveaways, phishing emails disguised as school-related messages, and clickbait links that install malware. Fraudulent messages may appear as notifications from gaming sites, social media platforms, or even impersonations of their friends. They may also try to trick children into revealing personal information by pretending to be authority figures or celebrities.

How to Navigate:

- **Teach children to be skeptical** of messages that promise free items or require immediate action.
- **Never click on unknown links** or download attachments from unfamiliar sources.
- **Enable Multi-Factor Authentication (MFA)** on all important accounts to add an extra layer of security. This can also be called Two-Factor Authentication (2FA).
- **Discuss common scams** with your child to help them recognize red flags. Examples include: free tickets or prizes or pretending to be a friend they don't remember.

Bonus: Run a scenario with your child where you pretend you are a scammer trying to get information from them. Have them act out how to spot it, and what they should do.

Risk 4: Dark Web & Dangerous Content

The dark web is an area of the internet not indexed by standard search engines, often associated with illegal activities, explicit content, and harmful forums.

Children can be exposed to harmful content unintentionally through search engine results, chat platforms, or misleading links. This can include violent imagery, self-harm encouragement forums, or extremist content.

How to Navigate:

- **Use safe search settings** on browsers, YouTube, and social media platforms to limit exposure.
- **Install parental control software** that filters out dangerous websites, see Chapter 3 for more details.
- **Keep an open dialogue** about internet curiosity and emphasize that not all information found online is safe or appropriate.
- **Teach your child to talk to you** if they encounter something disturbing online.

Bonus: Ask your child about some of the content they have found interesting while browsing. Open a dialogue for them to come to you about what they are learning, but also what they are seeing online.

What Is Still "In Your Control" vs. What Becomes "Out of Your Control"

The internet is vast, and while you can set rules and monitor your child's usage, some elements are beyond your control. No matter how many safety measures or quick actions you take, there will always be moments when your child encounters something you didn't foresee. But we can prepare them with education and help them develop how to react to items out of our control.

The digital world is constantly changing, with new apps, games, and social platforms emerging all the time. This means there will always be some "blind spots" – areas of risk you simply don't have visibility into. It's important to recognize that your role isn't to eliminate every risk, but rather to prepare your child with the knowledge and tools they need to navigate these unseen dangers confidently.

That's why **open communication is so critical**. You can't see everything your child does online, but you *can* equip them to make safer choices when they encounter something risky. In later chapters, we'll talk about the steps you can take to empower your child to protect themselves in situations where you may not have control.

Recognizing these boundaries will help you focus on what *you* can do to create a safer digital environment for your child.

In Your Control:

- Setting parental controls and privacy settings (see Chapter 3).
- Having open conversations about online risks.
- Teaching kids about online etiquette, digital footprints, and security.
- Monitoring their activity and ensuring they know they can talk to you.
- Enforcing device-free times and safe browsing habits.
- Leading by example with your own online habits.

Out of Your Control:

- Other people's actions online (cyberbullies, predators, scammers).
- The ever-changing privacy policies of social media platforms.
- Content algorithms pushing videos and posts your child may not be ready for.
- Data breaches exposing personal information.
- The permanence of shared content, including images and personal details.

Conclusion

As your child journeys through the ever-evolving stages of technology – from wide-eyed toddlers swiping at screens to teens navigating social media feeds – the role you play as a guide, protector, and teacher cannot be overstated.

Each phase comes with its own challenges, but also with opportunities to build trust, resilience, and digital wisdom. While it may feel overwhelming at times, remember that your presence and proactive involvement are powerful safeguards. With the right mix of boundaries, open conversations, and consistent check-ins, you're helping your child grow into a thoughtful, safe, and confident digital citizen.

Though the digital world is vast and constantly shifting, your influence remains a steady anchor. You will not be able to control every link they click or every message they receive, but you *can* prepare them for those moments. By focusing on what is within your control – your example, your conversations, and your ongoing guidance – you are giving your child the tools they'll need long after they've left the nest.

And just like learning to cross the street safely or ride a bike, navigating the digital world starts with trust, practice, and your steady hand at their side.

Next Steps for Parents

Understanding the risks is the first big step. Now let's put that knowledge into action – one small step at a time. Here are five practical things you can do today to start protecting your child in this digital world.

Step 1: Do a Digital Footprint Check
(10 Minutes)

- Search your child's name in Google, including images, to see what public info is out there.
- Review your own social media posts for any tagged photos or personal details about your child.

Bonus: Check your privacy settings on Facebook, Instagram, and Google Photos to ensure you're not unintentionally sharing more than intended. Some examples include your phone number and location tracking.

Step 2: Have a "Stranger Danger – Online Edition" Talk *(5 Minutes)*

- Explain that just like we don't talk to strangers in real life, the same rule applies online – even in games and group chats.
- Use real age-appropriate examples of how strangers might try to engage them online and what to do if it happens.

Bonus: Practice saying "no" or blocking someone together, so your child knows what it looks like in their favorite apps or games.

Step 3: Review Settings on Their Favorite Apps and Games *(15 Minutes)*

Understand what they're using and what those platforms expose them to.

- Go through the apps or games your child uses and review their chat features, privacy settings, and content filters.
- Adjust settings to limit who can contact them and disable location sharing where possible.

Bonus: Use this as a time to let your child teach you about their favorite platform – it makes the experience feel more like a conversation or bonding moment rather than surveillance.

Step 4: Create a Cyber Safety Family Rule *(5 Minutes)*

- Pick one safety rule together, such as "Never post personal info in public profiles" or "Ask a parent before downloading a new app."
- Write it down, decorate it, and put it somewhere visible like the fridge or family calendar.

Bonus: October is Cybersecurity Awareness Month, during which you can make a special flip book of all the rules you and your child have made during the year. This will help set norms and expectations for safe online behavior.

Step 5: Introduce the "Pause and Think" Rule *(2 Minutes)*

Beyond Think before you Post, this is a quick mindset shift that builds long-term habits specific to what is written online in chats, comments, or online forums.

- Teach your child to pause before posting or responding online: "Would I say this out loud in real life?"
- Reinforce that once something is online, it can be hard to take back – even if deleted.

Bonus: Model this yourself. Next time you're about to post something, talk aloud about how you decided it was okay to share.

Chapter 3

Introduction

Today's kids don't just use technology – they live in it. As parents and caregivers, it's easy to feel overwhelmed by the pace of new devices, apps, and platforms. But the truth is, guiding our children through the use of phones, tablets, game consoles, and more doesn't have to be complicated. It simply requires intention, consistency, and a willingness to learn alongside them.

This chapter is your hands-on guide to making the most of the devices your children use every day – while keeping them safe and helping them build strong digital habits. Whether it's setting parental controls on a smartphone, enabling safe search on a smart TV, or teaching the value of strong passwords and MFA, you'll find clear, current steps you can take right now. These tools aren't about surveillance – they're about support. When used with care and open communication, they create a tech environment that empowers your child to explore, learn, and grow with confidence.

As you move through the tips and checklists, remember: you're not expected to be perfect. Device settings will change. New threats will emerge. What matters most is that your child knows they can come to you – not just when something goes wrong, but as an ongoing part of their digital journey. Think of this chapter as a toolkit, and yourself as both the guide and the guardrail. With a little structure and a lot of heart, you can help your child build a relationship with technology that's safe, balanced, and rooted in trust.

Guiding and Monitoring Device Use – Phones, Tablets, and Beyond

Every device your child uses is a learning space, and like any good playground, a little structure helps everyone have more fun and stay safe.

Parental tools aren't about spying – they're about partnering. Below we have listed some commonly used devices and their current (2025, time of publication) instructions to help you stay informed and support healthy tech habits. Consider this as a starting point, but research your own device safety settings that may exist in your environment.

From time-to-time, operating systems are updated in a way that drastically changes how you access or modify the settings. When in doubt, use a search engine to figure out the steps for your system, platform, or application.

Smart TVs, Game Consoles, Laptops, and Network Devices:

1. For Smart TVs (e.g., Roku, Samsung):
   - Go to **Settings > Parental Controls**.
   - Set content PIN based on rating (e.g., TV-Y7, PG-13).
   - Netflix: Go to *Manage Profiles* > Set *Kids* profile.
   - YouTube: Use **YouTube Kids** app.
   - Disney+: Set up a *Kids Profile* and PIN for adult content.

33

2. For PlayStation:
   - Create a child account under your family group.
   - Go to *Settings* > *Parental Controls.*
   - Limit screen time, chat, and age-restricted games.

3. For Xbox:
   - Use the Xbox Family Settings app.
   - Set time limits and filter content.

4. For Nintendo Switch:
   - Download **Nintendo Parental Controls** app.
   - Set play-time limits and restrict mature content.

5. For Windows PCs:
   - Use **Microsoft Family Safety**.
   - Go to *Settings* > *Accounts* > *Family & other users.*
   - Add a child account, Set screen time limits, block inappropriate content. Get weekly activity reports via email.

6. For macOS:
   - Go to *System Preferences* > *Screen Time.*
   - Set limits for websites, apps, and usage.

7. Wi-Fi Routers with Parental Controls
    ○ Choose a router with built-in parental controls (e.g., Eero, Google Nest, Netgear).
    ○ Log into your router app.
    ○ Set bedtime schedules for devices.
    ○ Block inappropriate sites across the network.

Devices Your Child Uses Independently:

iPhone & iPad (as of iOS version 18.4.1)

1. Go to *Settings > Screen Time*.
2. Tap **Turn on Screen Time** and select *This is My Child's iPhone/iPad*.
3. Tap **Downtime** and schedule off-hours (e.g., 8 PM–7 AM).
4. Go to **App Limits** > Add limits for categories like Games, Social Media.
5. Set a **Screen Time Passcode** only you know.
6. Tap **Content & Privacy Restrictions** >
    a. Enable Restrictions such as web content, purchases, explicit music, and app ratings.
    b. Under **Account Changes**, set to *Don't Allow* to lock down settings.
7. Keep the conversation open as your child grows and update settings together.

Android Phone & Tablet Devices (as of Android version 15)

1. Download **Google Family Link** from the Play Store.
2. Create a Google Account for your child if they don't have one.
3. Link your Google Account to your child's via Family Link.
4. Use the app to:
   - Approve or block app downloads.
   - Set screen time limits per app.
   - Lock devices remotely.
   - View app activity reports.
5. Go to *Settings > Digital Wellbeing & Parental Controls*.
6. Set filters for apps, websites, and screen time.
7. Disable installing from unknown sources: *Settings > Security > Install unknown apps > Set to Not allowed.*

Devices You Use that Your Child Has Access to:

Sometimes you want your child to use an app – like a learning game or a drawing tool – without wandering off into other apps or settings. Luckily, both iOS and Android offer built-in tools to help. Here's how to keep your child's device to a single app.

iPhones & iPads – Turning on Guided Access:

Guided Access is Apple's solution for keeping your child inside one app and preventing them from tapping their way elsewhere.

To Turn On:
1. Open Settings
2. Scroll down and tap Accessibility
3. Scroll to the bottom and tap Guided Access (under "General").
4. Turn Guided Access on.
5. Set a Passcode

Tap Passcode Settings to create a special code (or use Face ID/Touch ID). This code is what you'll use to exit Guided Access mode.

To Use It:
1. Open the app you want your child to use. For example, open a coloring app.
2. Triple-click the side or lock button (at the time of publication in 2025, this is still a physical button).
3. A Guided Access screen will appear, tap Start.

Your child cannot leave the app, open other apps, or change settings.

4. To end it: Triple-click again, enter your passcode, and tap End.

Android Devices – Using Screen Pinning (App Pinning):

Most Android phones and tablets come with Screen Pinning, which works similarly to Guided Access by keeping the user in one app.

To Turn It On:
1. Open Settings
2. Go to Security or Security & Privacy (On some phones, it might be under Biometrics and Security.)
3. Scroll to the bottom and tap Advanced, then find Screen Pinning or App Pinning
4. Toggle it On. Also turn on the option that requires your PIN to unpin.

To Use It:
1. Open the app you want your child to use. For example, a math practice app.
2. Open the app switcher, swipe up from the bottom and hold, or tap the square/multitask button, depending on your device.
3. Tap the app's icon at the top of its preview and select Pin or Screen Pin. Screen Pinning doesn't let you block parts of the screen like Guided Access does, but it still prevents app-switching and keeps your child where they need to be.
4. To end it: Hold the back and overview buttons at the same time (or follow your phone's prompt), then enter your PIN.

By setting up Guided Access or Screen Pinning, you can give your child the freedom to explore an app, and you get peace of mind knowing they won't end up in your email or making mystery purchases.

Building Strong Accounts: Passwords & Multi-Factor Authentication (MFA)

Your child's digital identity is worth protecting. With just a few tweaks, their accounts can be safe from hackers, trolls, and mischief-makers.

Creating Strong Passwords

1. Avoid using names, birthdays, or common words. Instead use fun phrases like GiraffesDance2024!
2. Include uppercase, lowercase, numbers, and special characters.
3. Ensure each site has a unique password.

Using a Password Manager

Password Managers are applications that use complex algorithms that make it difficult for hackers to crack. It manages your passwords so that you don't have to remember them or use the same password across accounts and devices.

1. Choose a password manager: *1Password, Bitwarden, LastPass* are just some examples.

2. Set up your master password – this should be the strongest.
3. Let the app generate and store complex passwords.
4. Sync across your devices.
5. Teach older kids to use it for school and game logins.

Enabling MFA

1. Go to account settings of each service (email, school, banking).
2. Enable **MFA**.
3. Choose the method:
   - **Authenticator App** (preferred method): Download *Google Authenticator*, *Authy*, or *Microsoft Authenticator*.
   - **Text message** (only if authenticator isn't available).
4. Scan the QR code or enter the key manually.
5. Save backup codes securely (in your password manager).

Securing Your Child's Accounts

1. Help them set usernames that don't include birthdates or real names.
2. Set strong passwords for school and gaming accounts.
3. Enable MFA on accounts that support it.
4. Review account security settings every 3 months.

Teaching Kids Cyber Awareness

Create a culture of communication and curiosity. Instead of lectures, lead with curiosity and shared learning. When kids feel empowered, they make better decisions.

Start Conversations Early

1. Ask open-ended questions: "What's your favorite app? Why?"
2. Share age-appropriate news about online scams or cyberbullying.
3. Talk about online privacy like you talk about real-world safety.

Practice Role-Playing

1. **Scam Simulation:** Pretend to send a phishing message offering a fake prize.
    - Ask: "What would you do"
    - Teach: "Don't click. Always check with a parent."

2. **Suspicious Friend Request:** Pose as a stranger.
    - Ask: "Would you accept this friend?"
    - Teach: "If you don't know them in real life, don't connect."

3. **Inappropriate Content:** Explain what something inappropriate might look like in an age-appropriate manner.
   ○ Ask: "What would you do if you saw something scary?"
   ○ Teach: "Close it, tell an adult, and block/report."

Daily Cyber Habits to Teach

1. Always log out of apps when done.
2. Don't share personal info.
3. Don't click unknown links or pop-ups.
4. Ask permission before installing new apps or games.
5. Use nicknames in games instead of real names.

Encouraging Healthy Digital Habits – Screen Time & Device Boundaries

Good tech use is about balance, not bans. Below we have listed some ideas that may work for you and your family. Chose and adapt ones that set the right balance of boundaries for your family.

Build Healthy Routines

1. Use mealtimes and before bed as tech-free windows.
2. Schedule daily screen breaks – stretch, play outside.
3. Define weekend vs weekday screen time limits.

Use a Family Tech Contract

1. Make it collaborative and draft rules *together*.
2. Include:
   - Where and when devices can be used
   - What happens if rules are broken
3. Sign it and post it somewhere visible. Revisit regularly.

*See **Family Tech Contract** at the end of this book for a template to start with.*

House Rules to Try

1. No phones in bedrooms after 9 PM.
2. 30 minutes of screen-free play before screens.
3. One day a week tech-free (Tech-Free Tuesdays).

Conclusion

Guiding your child's use of devices isn't about perfect control – it's about creating a thoughtful, flexible framework that grows with them. From setting up parental controls to teaching them how to create strong passwords and recognize red flags, you're laying the groundwork for habits that will serve them for life. The goal isn't just to protect them in the moment, but to prepare them for the moments you won't be there to guide every click or swipe.

Remember, the best digital safety tools are built on strong relationships. Keep the conversation open, review settings regularly, and involve your child in the process so they feel empowered, not policed. As technology evolves, your guidance will help them stay curious, confident, and most importantly, safe.

Next Steps for Parents

You don't need to be a tech expert to create a safer digital world for your family. These quick, practical steps will help you turn today's tools into tomorrow's healthy habits.

Step 1: Set Up Screen Time or Family Link
(10 Minutes)

- Update your device using the steps in this chapter that will protect your children when they use it. Be aware of settings you put in place that may block you from using your phone after the child is done.

- Update devices specifically used by a child using the steps in this chapter to protect your child when they use it. A child specific device can be locked down much more than your own device.

Step 2: Strengthen Your Child's Logins
(10 Minutes)

Protect their digital identity like you would their house keys.

- Help them create strong, unique usernames and passwords for school and game accounts.
- Enable MFA on digital platforms.

Bonus: Try not to use personal information like name, birthdates, or pet names.

Step 3: Add Monitoring Tools You Trust
(8 Minutes)

Monitoring is about partnership, not policing.

- Try *Bark* or *Qustodio* to receive alerts about risky content and behavior.
- Review weekly activity summaries and adjust settings as your child grows.

Step 4: Talk Cyber Smarts During Downtime
(5 Minutes)

You don't need a lecture – just a conversation.

- Ask: "What's a weird message you've gotten online?"
- Share: "Here's how I spot fake messages or scams – want to try together?"

Step 5: Draft a Family Tech Contract
(15 Minutes)

- Include rules like "No phones in bedrooms after 9 PM" or "Ask before installing apps."
- Sign it together and hang it on the fridge. Revisit monthly or yearly as habits evolve.

*Bonus: See **Family Tech Contract** at the end of this book for a template to start with.*

Chapter 4:

PRIVACY & DATA PROTECTION for FAMILIES

Introduction

In today's world, personal data is as valuable as currency. Unfortunately, families are among the most vulnerable targets for cybercriminals, data brokers, and identity thieves. From unsecured home networks to children unknowingly exposing personal details online, the risks are everywhere.

Consider this: A 2023 report by the FTC found that children's personal data is increasingly targeted in cyberattacks, with school database breaches exposing sensitive student information, including Social Security Numbers and addresses. In one case, a hacker sold the Social Security Numbers of thousands of minors on the dark web, leading to years of identity fraud before many families even realized what had happened. These incidents highlight how critical it is for parents to take cybersecurity seriously.

In Chapter 1, we discussed what a digital footprint is. This chapter provides a practical guide to securing your family's digital footprint. But where do you start? We will work from the point you connect to the internet at home and on the go to how to safely use the internet. You'll learn how to protect your home Wi-Fi and devices, ensure your children browse the internet safely, and take steps to prevent identity theft – especially for minors, who are increasingly targeted.

How to Secure Your Home Wi-Fi & Devices

Your home network is the doorway to all your family's devices, and if it's not secure, hackers can find their way in. They might steal personal info, install harmful software, or spy on your activity.

Start with your router

Don't stick with the default settings from your internet provider. Change the admin password to something strong and unique. Use WPA3 (or WPA2) for Wi-Fi security and avoid outdated options like WEP. Keep your router's software (firmware) up to date to patch any security holes.

Smart devices like TVs, speakers, or baby monitors also need protection

Always change default passwords, update their software regularly, and turn off features you don't use – especially remote access. Set up a separate guest Wi-Fi for visitors to keep your main network safer. When buying smart devices, choose trusted brands; the cheap options aren't worth the risk if it puts your privacy in danger.

Mobile phones are another common target

Keep your phone and apps updated, only download apps from official stores, and check what permissions each app really needs. For example, does your app require your location or can that be turned off? Do you allow it to share your data and track your data with and across other apps? Uninstall apps you don't use. Use strong passwords, biometrics, or built-in password tools. Turn off Bluetooth when not in use and avoid open Wi-Fi networks that do not require a password.

Finally, protect your privacy

Turn off personalized ads, limit location tracking, and check which apps can access your camera, mic, or contacts. A few small settings can go a long way in keeping your data safe from unwanted tracking.

To further protect your privacy, you can also use encrypted messaging apps like Signal that uses end-to-end encryption by default, meaning only the sender and receiver can read the messages.

Safe Browsing & Search Engines for Kids

Children often don't recognize online threats, making them easy targets for scams, inappropriate content, and data collection.

Educating Children on Internet Safety

Talk to your kids about online risks. Teach them never to share personal information (full name, address, school name, birthday, pet names) online. Encourage open conversations about suspicious websites or messages.

Parental Controls & Filtering Tools

Use built-in controls on devices, browsers, and operating systems to filter content. Services like OpenDNS FamilyShield can block harmful sites at the network level.

On YouTube, enable Restricted Mode to filter out inappropriate content. Browser extensions like uBlock Origin can help block ads that lead to malicious sites.

Smart TVs offer parental controls to block content in applications like Disney+ and Netflix with a simple code that only parents know.

Privacy-Focused Search Engines and Browsers for Kids

For search engines, Google collects massive amounts of data, even in safe mode. Alternatives like Kiddle, DuckDuckGo, and Swisscows prioritize privacy and filter out inappropriate content.

When selecting a browser, check to see if there is a "Kids Mode" like in Microsoft Edge, which is a built-in-kid friendly mode that blocks content that children should not be exposed to. Other browsers like Safari and Chrome have built-in safe browsing modes also block content that is not age-appropriate.

Identity Theft Protection for Children

Children's Social Security Numbers and other personal information are valuable to identity thieves because they provide a clean slate for fraud.

Understanding the Risk of Child Identity Theft

Unlike adults, children don't check their credit reports, meaning fraud can go unnoticed for years. Schools, pediatrician offices, and even social media profiles can expose personal details that criminals exploit.

Proactive Steps to Protect Children's Personal Information

One of the best protections is freezing your child's credit with the major bureaus (Experian, Equifax, TransUnion). We mentioned this in Chapter 1 – if you haven't done this yet, take a moment to check that off your to-do list. This prevents anyone from opening accounts in their name. It is also easy to temporarily unfreeze these accounts when they reach an age where they need to establish credit.

Monitor your child's information for breaches by using services like Have I Been Pwned (haveibeenpwnd.com) or credit monitoring tools. Avoid oversharing on social media – never post full names, birthdates, or school details.

Spotting Possible Identity Theft

If you receive credit card offers in your child's name, it could be a red flag. Check their credit report for any fraudulent activity and report issues to the FTC and credit bureaus.

Conclusion

Protecting your family's data requires ongoing effort. Cyber threats evolve, and staying informed is key. By securing your home network, teaching children safe online habits, and proactively guarding against identity theft, you create a safer digital environment for your family.

For further resources, consider subscribing to cybersecurity blogs, setting Google Alerts for privacy breaches, and using tools like password managers and encrypted messaging apps.

Next Steps for Parents

Taking control of your family's digital security doesn't have to be overwhelming. Follow this step-by-step checklist to start implementing best practices today.

Step 1: Secure Your Home Wi-Fi *(5 Minutes)*

- Change your router's default admin password to a strong one with a mix of uppercase, lowercase, numbers, and symbols.
- Check your Wi-Fi encryption – set it to WPA3 (or WPA2 if WPA3 is unavailable).
- Update your router's firmware by logging into your router and checking for updates.

Bonus: If your router is more than five years old, consider upgrading to a newer model with built-in security features. This can include automatic firmware updates, ability to have guest network support, an interface to monitor connected devices, and more.

Step 2: Lock Down Smart Devices *(10 Minutes)*

- Change the default passwords on smart devices (TVs, cameras, baby monitors, etc.).
- Turn off remote access unless absolutely necessary.
- Set up a separate guest Wi-Fi for visitors and their devices.
- Update your operating system and applications frequently and set up auto updates. Be sure to check for privacy updates or changes with these updates.
- Limit tracking by disabling personalized ads and restricting application permissions, especially location, microphone, and camera.

Bonus: Check for software updates on your smart home devices and install them. This ensures your devices have the most recent security updates. Remember to check privacy settings after updates to ensure nothing has been reset.

Step 3: Make Internet Safety a Habit for Your Kids *(15 Minutes)*

- Have a quick chat with your kids about online dangers. Teach them:
    o Never share personal information (name, address, school, birthday, pet names).
    o Be cautious of links and messages from unknown people.
    o Always ask before downloading apps or clicking on links.
- Turn on parental controls on their devices and web browsers.
- Switch to a kid-friendly search engine (Kiddle or DuckDuckGo Kids).
- Stay informed about the apps and platforms your child uses.
- Use parental monitoring tools like Bark, Qustodio, or Google Family Link.

Bonus: Ask your child which websites and apps they like most, which opens the door for ongoing discussions.

Step 4: Protect Your Child's Identity *(10 Minutes)*

- Check if your child's information has been exposed using Have I Been Pwned (haveibeenpwnd.com)
- Stay vigilant while filling out school forms; avoid giving permission to store or post photos of your child.

Bonus: Set a reminder to check for your child's information online once a year.

Step 5: Strengthen Password Security
(10 Minutes)

- Use a password manager (e.g., 1Password, Bitwarden) to store and generate strong passwords.
- Enable MFA on all accounts, especially banking, email, and social media.

Cybersecurity doesn't have to be complicated. By following this step-by-step plan, you can significantly reduce your family's risk in under an hour.

Chapter 5

RESPONDING TO CYBER THREATS

Introduction

The internet offers incredible opportunities for learning, entertainment, and connection, but it also exposes children to serious cyber threats. If your child's data is breached, bad actors can assume their identity and cause long-lasting damage to their credit. Another major concern is cyberbullying, including online harassment. Parents must be prepared to respond effectively.

Ignoring these threats can have lasting psychological and emotional consequences. This chapter outlines practical steps to take if your child faces cyberbullying, how to report online threats, and when to involve law enforcement.

How to Respond if Your Child's Privacy is Breached Online

Assess the Situation

First, determine exactly what information has been exposed. Check whether personal details such as your child's name, address, social media accounts, or financial information have been leaked. If images or videos have been posted without consent, identify where they are being shared and by whom. Take screenshots of all concerning content before attempting to remove it, as this evidence may be needed for reporting or legal action.

Secure Your Child's Accounts

If your child's accounts have been compromised, take immediate action to regain control. Change all passwords, starting with the affected platforms, and ensure that each account uses a strong, unique password. Enable MFA for added security. If a hacker has gained access, log out of all devices and re-sign in to force unauthorized users out.

Report and Request Content Removal

Once you have documented the breach, start the process of removing the content. Use reporting tools on social media platforms to flag and request the removal of unauthorized photos, videos, or personal information. If the content appears on a website, contact the site administrator for a takedown request. The Digital Millennium Copyright Act (DMCA) may apply if images or content were posted without consent.

For search engines like Google, use their content removal tool, especially if the content is explicit or harmful. If the breach involves harassment, exploitation, or identity theft, report the incident to local law enforcement and the Internet Crime Complaint Center (IC3).

Monitor for Identity Theft

If sensitive personal information, such as a Social Security Number, has been exposed, take steps to prevent fraud. Freeze your child's credit with the three major credit bureaus – Experian, Equifax, and TransUnion – to prevent unauthorized accounts from being opened in their name.

Regularly monitor for suspicious financial activity, such as fraudulent charges or new credit applications. If identity theft has already occurred, file a report with the FTC at IdentityTheft.gov to begin the recovery process.

Educate and Prevent Future Incidents

To protect your child from future privacy breaches, educate them about online safety and responsible digital behavior. Teach them the importance of keeping personal details private and recognizing potential online threats. Regularly update privacy settings on their social media accounts and devices to limit exposure.

Consider using parental controls and tools to keep informed about online activity. Staying informed about emerging cyber threats and taking proactive security measures will reduce the chances of another breach. If cyberbullying or online harassment is involved, seek support from legal professionals, school counselors, or child advocacy groups.

What to Do if Your Child is Cyberbullied

Recognizing Cyberbullying

Cyberbullying can have severe psychological, emotional, and even physical effects on children. It can lead to anxiety, depression, low self-esteem, academic struggles, and, in extreme cases, self-harm or suicide.

To illustrate this, consider the following hypothetical case of "Laura," a fictional character:

Laura, a 13-year-old girl, was an active social media user. One day, a classmate took an unflattering photo of her without her consent and posted it online with cruel comments. The post rapidly spread online, and classmates started sending her hateful messages. The relentless cyberbullying caused Laura to withdraw from friends, her grades dropped, and she developed severe anxiety. She avoided school, had trouble sleeping, and ultimately needed therapy to recover.

Cyberbullying takes many forms, including:

- **Harassment:** Repeated harmful messages, insults, or threats.
- **Exclusion:** Intentionally leaving someone out of online groups or chats.
- **Impersonation:** Creating a fake account to post harmful content or spread lies.
- **Doxxing:** Publicly sharing personal information like phone numbers or addresses.

Warning Signs Your Child May Be a Victim:

- Avoiding their phone or social media
- Emotional distress after being online
- Withdrawal from social activities or friends
- Changes in sleep or appetite
- Hesitation to discuss online interactions

How to Support Your Child

Children often fear that telling a parent will make things worse. The first step is listening without judgment. Parents must trust their children but also verify with other adults involved.

- **Validate their feelings.** Let them know it's not their fault.
- **Keep a record.** Take screenshots and note timestamps – this is essential if further action is needed.
- **Discuss options calmly.** Avoid reacting with immediate anger or retaliation.

Taking Action

- **Block the bully.** Teach your child how to block users on social media, mobile devices, and gaming platforms.
- **Report the behavior.** Most platforms have policies against harassment.
- **Adjust privacy settings.** Review and tighten security on accounts.
- **Involve the school (if applicable).** Many schools have policies against cyberbullying, even if it happens outside school hours.

How to Report Online Threats & Harassment

When and Where to Report

Many online platforms have reporting tools for cyberbullying and harassment.

- **Facebook & Instagram:** Report harassment via Settings → Help Center.
- **Snapchat:** Use the in-app reporting function under "More Options."
- **Gaming Platforms:** Roblox, Fortnite, and Minecraft allow reporting of in-game harassment.
- **CyberTipline (NCMEC):** Report online child exploitation at cybertipline.org.
- **StopBullying.gov:** Federal resources for cyberbullying support.

Preserving Digital Evidence

Bullies often delete posts or messages to erase their tracks. Before reporting:

- **Take screenshots** of messages, usernames, and timestamps.
- **Note details** of how often incidents occur and if they escalate.
- **Store evidence** in a digital folder for authorities if needed.

Understanding Platform Policies

Social media companies often take action against harassment, but enforcement can be inconsistent. If a report isn't addressed:

- **Appeal** within the platform's help center.
- **Escalate** by contacting their legal team or safety department.
- **Warn other parents** in private groups or forums.

When to Seek Law Enforcement Help

Situations that Warrant Legal Action

Not all online harassment requires police involvement, but some situations cross the line into criminal activity. Contact law enforcement if:

- Your child receives threats of violence.
- Someone shares explicit images or requests them from your child.
- A stranger stalks, follows, or manipulates your child online.
- Your child's identity is stolen or used fraudulently.

How to File a Police Report

1. Gather all evidence – Screenshots, emails, texts, and profiles.

2. Identify the jurisdiction – Cybercrimes can be reported to local police, but some cases (like child exploitation) should go directly to the FBI or specialized agencies.
3. Contact the right authorities:
   o Local Police – for immediate threats and harassment.
   o FBI Internet Crime Complaint Center (IC3) – for major cybercrimes: ic3.gov
   o National Center for Missing & Exploited Children (NCMEC) – for cases involving minors: missingkids.org

Legal Protections for Minors

Cyberstalking and harassment laws vary by state and country, but most have laws protecting minors from online threats.

- **Cyberstalking Laws:** In most places, repeated online threats can lead to restraining orders or criminal charges.
- **Sextortion & Online Exploitation:** Sharing explicit content of a minor is a federal crime.
- **Parental Consent Protections:** The Children's Online Privacy Protection Act (COPPA) limits how websites collect and use kids' data.

An example of a law that was put in place in 1994 and updated in 2019 to align with generative AI and the rise of deep fakes – In Virginia, it is illegal to create images of others without their consent under specific circumstances, as outlined in §18.2-386.1 of the Code of Virginia, titled "Unlawful creation of image of another; penalty."

Those who are found guilty of breaking this law in Virginia could expect a Class 1 Misdemeanor: A general violation of this statute is classified as a Class 1 misdemeanor or even a Class 6 Felony if the non-consenting person is under 18 years of age.

This is something as parents we should be aware of and educate our children who are often not aware of the damage they can do to others or their own lives by posting deep fake media.

Tip: Search your state to see if there are applicable laws that you should be aware of

Conclusion

Protecting children from cyber threats requires vigilance, education, and proactive action. Online privacy breaches can lead to identity theft, financial fraud, and emotional distress, making it essential for parents to secure accounts, monitor for suspicious activity, and report violations.

Cyberbullying, which can severely impact a child's mental health, must be addressed with a supportive approach, clear documentation, and swift intervention. Knowing when and how to report online threats – whether to platforms, schools, and law enforcement – ensures that harmful behavior is properly handled.

By staying informed, enforcing strong security measures, and fostering open communication, parents can help their children navigate the digital world safely and confidently.

Next Steps for Parents:

Cyber threats are an unfortunate reality, but parents can take strong, proactive steps to protect their children. Use this checklist to begin building a safer digital environment at home.

Step 1: Lock Down Devices and Apps
(20 minutes)

- Set up appropriate parental controls on phones, tablets, and computers.
- Adjust privacy settings on social media, gaming platforms, and school apps.

Bonus: Disable location sharing unless absolutely necessary.

Step 2: Keep Lines of Communication Open
(30 minutes)

- Explain what to do if they receive a suspicious message or see something upsetting.
- Reassure them they can always come to you without fear of punishment.

Bonus: Have a conversation tailored to your child's age about personal information, digital boundaries, and online respect. *See **Talking Points** at the end of this book.*

Step 3: Celebrate Safe Digital Habits
(5 Minutes)

- Praise your child for reporting something suspicious or sticking to screen limits.
- Let them pick the next tech-free family activity – ice cream outing, movie night, or board game time.

Let these small wins stack up. You're not just setting limits – you're setting them up for success.

Step 4: Monitor and Adjust Regularly
(Ongoing)

- Check in with your child about their digital life.
- Review new apps or websites together before allowing access.

Bonus: Stay informed about new online trends and threats via news feeds, online forums, and your own research.

By staying informed and taking action early, you're not just reacting to problems – you're creating a safer digital environment for your child and empowering them to navigate online spaces responsibly.

Talking Points

Stick to the basics and practice the following throughout your child's upbringing:

Kindergarten to 5th Grade (Ages 5-10)

ONLINE SAFETY TIPS

NOT EVERYTHING SHOULD BE SHARED

Full names, addresses, phone numbers, school names, and photos are secret info!

ASK BEFORE YOU CLICK OR SHARE

Always ask a parent before downloading apps, clicking links, or chatting online.

STRANGERS ONLINE ARE STILL STRANGERS

People in games or chats are not always special info friends.

ALWAYS CHECK WITH A PARENT

Get permission before giving info to a website or app

Key Goals: Introduce basic concepts of privacy and safe behavior online.

Discussion Points:

- **"Not everything should be shared."** Explain that full names, addresses, phone numbers, school names, and photos should only be shared with family or trusted adults.

- **"Ask before you click or share."** Teach them to ask a parent before downloading apps, clicking links, or chatting online.

- **"Strangers online are still strangers."** Even if someone seems friendly in a game or chat, they shouldn't be trusted with personal information.

- **"Always check with a parent"** before giving information to a website or app.

- **Use kid-friendly terms** like "secret info" or "special info" to describe private data.

WHAT YOU POST STAYS ONLINE

Even deleted posts, photos, or messages may still be saved or shared by others

THINK BEFORE YOU SHARE

Pause and consider how a post or photo might be seen now and in ine future

PRIVATE INFO INCLUDES MORE THAN JUST YOUR NAME

That means less obvious details, like school location or hobbies

REVIEW YOUR PRIVACY SETTINGS

Keep up-to-date to help limit who can see your profile

Key Goals: Build awareness of privacy settings, digital footprints, and social risks.

Discussion Points:

- **"What you post stays online."** Teach them that even deleted posts, photos, or messages may still be saved or shared by others.

- **"Think before you share."** Encourage them to pause and consider how a post or photo might be seen by others – now and in the future.

- **"Private info includes more than just your name."** Talk about less obvious details like school location, hobbies, or where they hang out.

- **"Review your Privacy Settings."** Help them review app and platform settings regularly to keep accounts locked down.

- **Online friends aren't always who they say they are.** Reinforce that they should never meet an online contact in person or share real-world details.

KEY GOALS:

Help them understand long-term consequences, data tracking, and their digital reputation.

- 'Nothing online is ever 100% private.' Even private messages and DMs can be screenshot or leaked.

- 'Be aware of online predators.' Talk openly about manipulation, flattery, and pressure that may be used by

- Protect logins and devices, Stress the importance of strong passwords, not sharing accounts, and enabling MFA authentication

- You're building a digital reputation. Colleges, jobs, and others may see what's posted publicly -- encourage them to curate a smart, respectful online presence.

Key Goals: Help them understand long-term consequences, data tracking, and their digital reputation.

Discussion Points:

- **"Nothing online is ever 100% private."** Even private messages and DMs can be screenshot or leaked.

- **"Free apps aren't really free."** Talk about how apps and platforms collect and sell data for advertising.

- **Protect logins and devices.** Stress the importance of strong passwords, not sharing accounts, and enabling MFA authentication.

- **"Be aware of online predators."** Talk openly about manipulation, flattery, and pressure that may be used by predators online.

- **You're building a digital reputation.** Colleges, jobs, and others may see what's posted publicly – encourage them to curate a smart, respectful online presence.

Tear the next page out and customize for your family.

Family Tech Contract

Purpose:

To create healthy habits for screen time, promote responsible online behavior, and keep everyone safe in our digital world.

1. Screen Time Rules

- I will follow the family limits for daily screen time:
  - Weekdays: _____ minutes
  - Weekends: _____ minutes
- I will take breaks every _____ minutes/hours to rest my eyes, move my body, and do something offline.
- Screens are off during:
  - Meals
  - Homework time
  - One hour before bedtime
  - Family activities
- I understand that too much screen time may lead to losing screen privileges.

2. Device Use

- I will ask permission before downloading apps, games, or joining any websites.
- Devices will be used in shared spaces – not in bedrooms behind closed doors.
- I will charge my device in the family charging station overnight.
- I understand that parents can check my device or activity at any time.

3. Online Safety

- I will never share personal information (full name, address, phone number, school, passwords, photos) without permission.
- I will only connect online with people I know in real life.
- If someone makes me feel uncomfortable or asks personal questions, I will tell a trusted adult immediately.
- I will never send, request, or forward inappropriate photos or messages.
- I understand that anything I post can be copied, saved, or shared – even if I delete it.

4. Respectful Behavior

- I will treat others kindly and respectfully online, just like I would in person.
- I will not say anything online that I wouldn't say face-to-face.
- I will not participate in or encourage cyberbullying, gossip, or exclusion.
- If I see something mean or harmful online, I will report it and tell an adult.

5. Consequences

- If I break this agreement, I understand that I may lose device or internet privileges for a set period of time.
- We will talk as a family to learn from mistakes and make better choices moving forward.

Child: _____   **Date:** _____

Parent/Guardian: _____   **Date:** _____

Reviewed Together On: _____ **Next Review:** _____

Glossary

Some items explained in more detail:

* **2FA (Two-Factor Authentication):** A security feature that requires two steps to log in – like a password plus a code sent to your phone.

* **AI Chatbots & Virtual Assistants:** Technology that answers questions, chats, or helps with tasks using artificial intelligence. Examples: ChatGPT, Google Bard, Alexa.

* **Anonymous & Encrypted Chat Apps:** Messaging apps that hide identities or use encryption to keep chats private. Examples: Telegram, Kik, Signal, Omegle.

* **COPPA (Children's Online Privacy Protection Act):** A federal law that imposes certain requirements on operators of websites or online services directed to children under 13 years of age, and on operators of other websites or online services that have actual knowledge that they are collecting personal information online from a child under 13 years of age.

* **Cyberbullying:** Bullying that takes place over digital devices like cell phones, computers, and tablets. Cyberbullying can occur through SMS, Text, and apps, or online in social media, forums, or gaming where people can view, participate in, or share content.

* **Cyberstalking:** The use of electronic communication to harass or frighten someone, for example by sending threatening emails or posting defamatory statements on social media.

* **Dark Web:** A part of the internet that is intentionally hidden and not accessible through standard web browsers. It is often associated with illegal activities.

- **Data Breaches:** Security incidents where sensitive, protected, or confidential data is copied, transmitted, viewed, stolen, or used by an unauthorized individual
- **Deep Fakes:** Synthetic media (audio, picture, video) in which a person in an existing image or video is replaced with someone else's likeness.
- **Doxxing:** Publicly revealing private personal information about an individual or organization, typically with malicious intent.
- **Educational Tools & Virtual Learning:** Online platforms schools use for remote classes, homework, and communication. Examples: Google Classroom, Zoom, Canvas, PowerSchool.
- **Grooming (Online):** The process used by predators to build a relationship, trust, and emotional connection with a child or young person so they can manipulate, exploit, and abuse them.
- **IC3 (Internet Crime Complaint Center):** A website run by the FBI that allows individuals to report internet crimes, including fraud, identity theft, and cyberattacks.
- **Multi-Factor Authentication (MFA):** An authentication method that requires the user to provide two or more verification factors to gain access to an online.
- **NCMEC (National Center for Missing & Exploited Children):** A private, nonprofit organization that operates the CyberTipline, a national reporting mechanism for online child sexual exploitation.
- **Online Gaming:** Games that are played over the internet and often involve chatting or playing with others. Examples: Roblox, Fortnite, Minecraft

- **Online Marketplaces & Shopping Apps:** Apps and websites where users can shop for goods, digital content, or services. Examples: Amazon, Etsy, Steam.

- **OpenDNS FamilyShield:** A free internet safety tool that blocks adult content and unsafe websites at the Wi-Fi level – no app required.

- **Phishing Scams:** Deceptive attempts to obtain sensitive information such as usernames, passwords, and credit card details by disguising oneself as a trustworthy entity in electronic communication.

- **Ransomware Attacks:** A type of malicious software (malware) designed to block access to a computer system until a sum of money (ransom) is paid.

- **Sexting:** Sending, receiving, or forwarding sexually explicit messages, photos, or videos, primarily via mobile phones or other digital devices.

- **Social Media:** Apps and websites where people share content and message each other. Examples: TikTok, Instagram, Snapchat, Discord.

- **Streaming & Entertainment:** Apps where users watch videos, shows, or live streams for fun. Examples: YouTube, Netflix, Twitch, TikTok Live.

- **Student Information Systems (SIS):** A digital platform schools use to manage student records, grades, attendance, schedules, and communication with families.

- **The Digital Millennium Copyright Act (DMCA):** A U.S. law that protects digital copyright – used to take down pirated or stolen content online.

- **WEP (Wired Equivalent Privacy):** An old and less secure wireless network security standard.

- **WPA2/WPA3 (Wi-Fi Protected Access):** Current and more secure wireless network security standards.

Appendix

Some additional details to help with the basics of online searching to help you continue your education as technology evolves over time.

Quick Guide: How to Use a Search Engine to Find Answers About Technology or Anything Else

Step 1: Open a Web Browser

- Look for an icon on your screen labeled Google Chrome, Safari, Firefox, or Microsoft Edge.

Step 2: Go to Google

- In the white bar at the top of the screen (called the address bar), type: www.google.com
- Press the Enter key (on a keyboard) or tap Go (on a phone or tablet).

Step 3: Type Your Question

- The white box in the middle of the page is the search bar.
- Click or tap in the box and type your question as clearly as you can.

Step 4: Press Search

- After typing your question, press the Enter key or tap the Search button (a little magnifying glass).

Step 5: Review the Results

- Google will show you a list of blue links and short summaries underneath them.
- Look for websites that sound trustworthy or familiar, such as:
  - Experts recommend not clicking on sponsored links, they may be fake or compromised websites.

Step 6: Click a Link to Read More

- Click or tap a blue link to open the full article or help page.
- Read through the page to find your answer. Many have pictures or step-by-step instructions.

Helpful Tips

- Use clear and simple words when searching.
- Add the name of the device or app you're using for better results (e.g., "iPhone parental controls").